iScience
Readers

Solar Energy:
Saving the School Budget

by Emily Sohn and Patricia Ohlenroth

Chief Content Consultant
Edward Rock
Associate Executive Director, National Science Teachers Association

NORWOOD HOUSE PRESS
Chicago, Illinois

Norwood House Press
PO Box 316598
Chicago, IL 60631

For information regarding Norwood House Press, please visit our website at
www.norwoodhousepress.com or call 866-565-2900.

Special thanks to: Amanda Jones, Amy Karasick, Alanna Mertens, Terrence Young, Jr.

Editors: Jessica McCulloch, Michelle Parsons, Diane Hinckley
Designer: Daniel M. Greene
Production Management: Victory Productions, Inc.

Paperback ISBN: 978-1-60357-314-6

Printed in Heshan City, Guangdong, China.
190P—082011.

CONTENTS

Note to Caregivers:

Throughout this book, many questions are posed to the reader. Some are open-ended and ask what the reader thinks. Discuss these questions with your child and guide him or her in thinking through the possible answers and outcomes. There are also questions posed which have a specific answer. Encourage your child to read through the text to determine the correct answer. Most importantly, encourage answers grounded in reality while also allowing imaginations to soar. Information to help support you as you share the book with your child is provided in the back in the **Additional Notes** section.

Words that are **bolded** are defined in the glossary in the back of the book.

Catching the Light

Every day, the Sun rises in the east and sets in the west. You may feel its heat and see its light. But chances are, you ignore it, at least most of the time. It's easy to forget something that's just always there.

Now may be the time to give the Sun your attention. After all, we wouldn't be here without it. The Sun's energy fuels weather patterns and keeps life ticking. **Solar energy** has a lot of practical uses, too. People use it to heat houses, make **electricity,** and more. In this book, you will learn how people capture and use solar energy. You will also solve a puzzle that challenges you to build a solar farm.

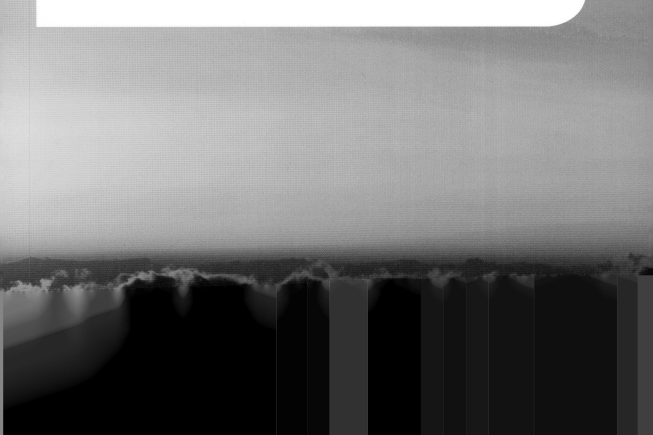

Farming the Sun

Your school has a problem. It is paying too much for electricity. If it can't cut its energy bills, the school board will have to cut costs and reduce services. This could mean larger class sizes, fewer after-school activities, or even staff reductions. You want to help and you have an idea: Turn energy from the Sun into light and heat energy for your school. First, you will need to build a solar farm. A solar farm doesn't have animals. Instead, a solar farm collects, stores, and delivers solar energy.

How are you going to help your school build a solar farm?

Before you get started, consider these questions:
• How will you capture solar energy?
• Where should you build the farm?
• What should the farm look like?
• What are the benefits of solar farms?

Draw some pictures of how you think your solar farm might look. Think about what materials you will use. Describe where you will put it. And explain how it will capture as much energy from the Sun as possible. Not sure how to address all these issues? You'll have a chance to revise your plan as you learn more.

To begin, try the Discover Activity to see solar energy at work.

When people think of a farm, they usually think of animals and barns.

Laser Focus

The family hadn't left any candles burning. So they were surprised when they returned to their house in Bellevue, Washington, one day in 2009. Their deck had mysteriously caught on fire. Investigators blamed the Sun. The family had left a glass water bowl on a wire rack in a sunny spot. Like a magnifying glass, they said, the bowl and the water could have **concentrated** the sunlight onto the deck. The energy may have been intense enough to burn wood.

Leaving a goldfish in a bowl of water outside in the sunlight is not a good idea. Can you explain why?

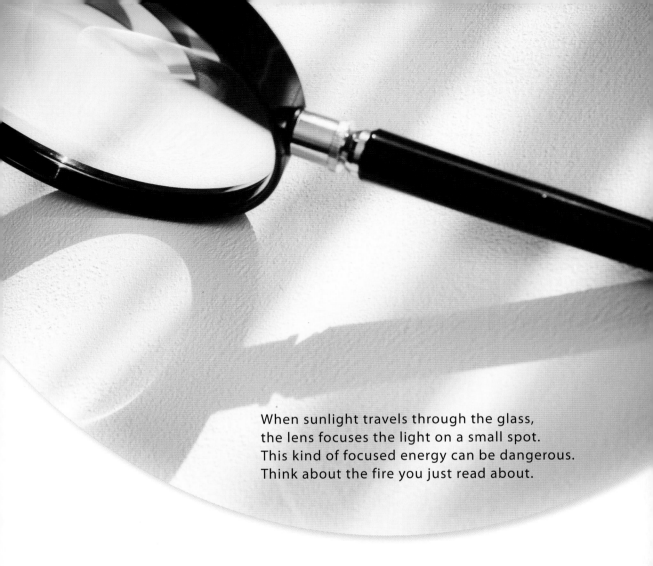

When sunlight travels through the glass,
the lens focuses the light on a small spot.
This kind of focused energy can be dangerous.
Think about the fire you just read about.

The Sun's energy spreads out as it travels through space to Earth. This energy comes in many forms, which include light and heat. By focusing sunlight with lenses or other tools onto a small area, you can concentrate that energy. Try the following activity. You'll learn how you can focus the Sun's energy.

How might concentrated sunlight help on your solar farm?

Materials
- aluminum foil
- black cloth (any material)
- 3 glasses
- water
- sand
- waterproof thermometer
- 3 white bowls
- measuring cup

white bowls

Line the inside of one bowl with black cloth. Line the inside of another bowl with foil. Make sure the shiny side of the foil faces up. Do not line the third bowl. Place all three bowls outside in front of a white wall when the Sun is high in the sky. If you cannot find a white wall, a light-colored wall will work. The bowls should be in direct sunlight.

Put a cup (8 ounces) of water into all three glasses. Place each glass in one of the bowls. Measure the temperature of water in each glass. Write down the numbers. Make sure to keep track of which temperature comes from which bowl. Now let the bowls sit out in sunlight for three hours. Measure and record the water temperatures again.

Look at your results. In each glass, did the water get warmer, colder, or stay the same? How do the temperatures compare from one glass to another? How might different colors and materials help you collect solar energy?

Did you find different temperatures in different glasses of water?

What Is Solar Energy?

Picture yourself by a lake on a sunny summer's day. You see light sparkling off the water. Your skin feels warm. The Sun gives us energy in the form of light and heat. It also gives plants the energy they use to make their food. Then, we eat plants as well as animals that eat plants. In other words, solar energy gives us the food we need for life.

Solar energy could be the source of some of the electricity in these wires.

The Sun's energy is plentiful. Some people use it to heat their homes. It can also be turned into electricity.

You studied solar energy in the Discover Activity. What did you learn?

Luckily for us, the Sun provides the right amount of energy to Earth to sustain life as we know it.

Earth

Our Star

Before you start tapping into solar energy, first learn about the big, bright star called the Sun. Yes, the Sun is a star. It's not the biggest star in the sky, but it's the closest star to Earth. It lies just 93 million miles (150 million kilometers) away. Earth is one of eight planets that orbit around it.

Like other stars, the Sun is a massive ball of **hydrogen** gas. Inside, **atoms** of hydrogen combine to form **helium** gas. The process is called **nuclear fusion.** Nuclear fusion releases tremendous amounts of energy.

Shadowed areas get less of the Sun's energy than sunny areas do.

Only a tiny amount of the Sun's energy reaches Earth. When that energy hits our atmosphere, clouds and water **reflect** some of it away. Sunlight also gets blocked and deflected on the ground. You make a shadow when sunlight hits you. How might shadows and reflections affect your solar farm?

Solar Energy at Work

We can't always see solar energy. But it's always there. According to a basic law of science, energy can't be destroyed and it can't be created. However, it can be stored. And it can change from one form to another. Solar energy is just one of many kinds of energy.

Energy from the Sun is responsible for the gusts you feel on a windy day.

Solar energy powers the weather. It heats land, air, and water. It causes liquid water to **evaporate** into gas. Water vapor rises and condenses into clouds. Eventually, it falls as rain or snow. This process is part of the **water cycle.**

The Sun also drives wind. When pockets of air get heated by the Sun, they move upward. Cooler air moves in to take their place. As a result, breezes blow.

The grass stores energy that will be passed on to the cattle, and from there, to the people who eat their beef.

Fed by the Sun

Without the Sun, life would probably not exist on Earth. Plants use energy from the Sun to make their own food. This is called **photosynthesis.** In the process, they release the oxygen that animals (including humans) need to breathe. When animals munch on leaves, seeds, and fruits, they are eating stored solar energy.

Animals use some of this energy to breathe and move. They store some, too. Meat contains energy that came first from the Sun.

Burning fossil fuels releases pollution into the environment.

The Sun feeds more than just life forms. It also helps create **fossil fuels.** These are materials that occur naturally on Earth. Today, people use them to make things work. Fossil fuels come in three forms: coal, oil, and natural gas.

Fossil fuels form over hundreds of millions of years. The process starts when plants and animals die. Some sink to the bottom of the sea or get covered in dirt. Over time, layers of dirt and rock bury them deeper and deeper. Then, heat and pressure inside Earth turn the remains into coal, oil, or gas. Fossil fuels are full of very old solar energy.

Today, we burn fossil fuels to heat homes, power cars, generate electricity, and make plastics. Do you think the supply of fossil fuels might run out some day?

How Do We Harness Solar Energy?

Fossil fuels store a lot of energy. Burning them creates heat energy and light energy. It also generates gases and other waste products that can pollute our air. Solar energy is much cleaner to use. Before you can use the Sun's rays, though, you have to catch them. One way to trap solar energy is with **solar collectors.** What these devices look like will depend on what you want to use the Sun's energy for.

The panels on the roof supply power for the house.
What do you think all the panels on the ground do?

Think about all the things you use energy for in your school. How will the energy you produce on your solar farm be used?

Most greenhouses use passive solar heating. If the temperature inside gets too warm, opening windows allows some heat to escape.

Soaking It Up

One way to use solar energy is to let the Sun do all the work. This process is called **passive solar heating.** No special equipment is necessary.

A passive solar building might have a lot of windows. They would let in as much sunlight as possible. Even on a cold day, you can feel warm in a sunny window. Trapped energy heats the floors, walls, and air. Warmed air then travels to cooler rooms.

Look on the Internet for other passive ways to warm or cool a building. How could you use passive solar heating to help your school?

Do you think a light-colored house like this one would be better in a warm climate or in a cool one?

There are ways to give passive systems a boost. In the Discover Activity, you should have seen that dark colors are best at soaking up heat. On a sunny day, that's why you feel hotter in a black shirt than you would feel in a white one. Other materials reflect light. White surfaces, shiny metals, and mirrors are a few. Solar energy bounces off these materials instead of soaking in.

How could you use dark and light materials on your solar farm?

This active solar heating system heats up water that moves through the solar collectors.

Taking Charge

Sometimes, passive solar heating systems aren't enough to meet a project's needs. In these cases, **active solar heating** can help. Special equipment helps collect energy.

Some solar collectors are flat panels. People put these dark plates on their roofs. The panels **absorb** energy and get warmer. Inside the panels, air or water gets warmer, too. The warmed air or water then moves into the house's heating system. There, it can heat up the whole house.

close-up of beach sand

Charging Up

Look around your school for all of the things that are plugged into a wall. You might find light bulbs, TVs, computers, and more. They all use electricity, which costs money to generate. One way to spend less on electricity is to use less. Another is to unplug electrical devices that are not in use. You can also make your own electricity.

Solar cells turn solar energy into electrical energy. Some calculators use solar cells. So do some electric cars. Satellites and spacecraft use them, too.

A solar cell looks like a thin wafer. It is made from a metal called **silicon.** Silicon is the second most common element in Earth's crust, or shell. You might see it at the beach. Combined with oxygen, it makes up glittering specks in the sand. Computer chips are also made from silicon.

How could solar cells be useful on your solar farm?

It's Electric

Remember, your task is to help your school save money on its energy bills. So you might want to make electricity on your farm. To do this, you should know how electricity works.

Electricity begins with the movement of electrons. These tiny particles are parts of an atom. They have negative charges. The path electrons follow is called an **electric circuit.**

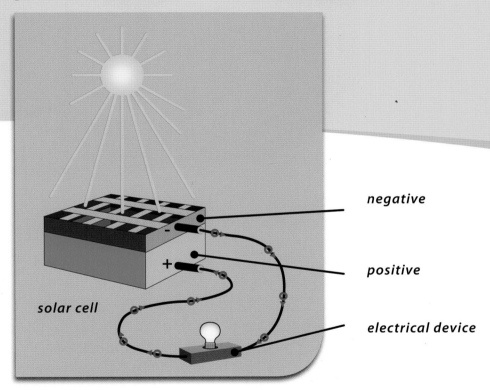

negative

positive

solar cell

electrical device

When sunlight hits a solar cell, it energizes electrons in the silicon atoms, knocking the electrons loose. Then the electrons move to the surface of the solar cell. There, a wire captures them. In a wire, electrons jump from one atom to the next. This movement is called an **electric current.** Currents are used to power electrical devices. Another wire returns the electrons to the back surface of the solar cell. That completes the circuit.

One toaster requires the equivalent of 800 solar cells to produce enough energy to heat your bread, turning it into toast for breakfast!

Solar cells can vary in size. Each area of a solar cell or panel can make only a certain amount of electricity. The amount of electrical energy is measured in **watts.** By some estimates, every square inch of a typical cell can produce 70 milliwatts of energy. (There are 1,000 milliwatts in a watt.) A clock radio needs 10 watts to run for 46 hours. A laptop computer needs 21 watts to run for 22 hours. To figure out how much energy you need, you must think about how many hours each electronic device will be running.

How many watts do you think it would take to power a whole school? How many solar cells do you think your solar farm would need? Keep in mind that you can get energy from more than one source. You might decide, for example, that you will get half of the energy you need from solar energy.

For what purpose might people use a large solar array like this one?

Power in Groups

To produce all the energy you need for your school, you'll need to connect a bunch of solar cells together. Groups of cells make up a **solar module.** Groups of modules form a **solar array.** One array can power the lights and appliances of a whole building. It's okay if it's not sunny all the time. Batteries can store solar energy for later use.

Do you want to use arrays on your solar farm? They can take up a lot of space. And arrays need to face the Sun. Think about your school. Where could you put an array of solar modules?

This glass marble concentrates the Sun's energy.

You don't have to use solar cells on your farm. Another kind of system uses mirrors. Mirrors focus sunlight into a beam. The beam's energy is used to heat a liquid. The hot liquid then powers engines. The engines in turn run electric generators. Mirror systems are cheaper than solar arrays, but they take up more space. And they are better for converting solar energy on a large scale, so they aren't as practical for just one home or building.

Many technologies exist for turning sunlight into energy we can use. Nobody knows yet which way will work best. Talk to your teacher or do some research. Which kind of methods would you want to use in your system?

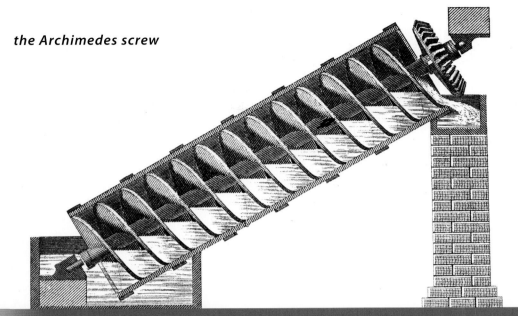

the Archimedes screw

CONNECTING TO HISTORY

Ancient War Machine

Archimedes was born in 287 BCE on the island of Sicily, off the coast of Italy. The Greeks ruled the island at the time. Archimedes became a great mathematician and scientist. He invented the Archimedes screw, a device that raised water from a lower level to a higher level. The Archimedes screw was used in irrigation.

During Archimedes' life, the Romans invaded Sicily. According to an old story, Archimedes used mirrors to concentrate sunlight on the invading ships. The energy from the focused light set the ships on fire. The Romans fled.

Some people think this story about Archimedes is true. Others are not convinced. What do you think? How could you find out more?

Solar collectors could gather all the energy we need if it were sunny all the time. These systems work best when pointed directly at the Sun. But clouds and pollution often block light from reaching solar cells. So does dust. Shade makes solar cells less efficient, too.

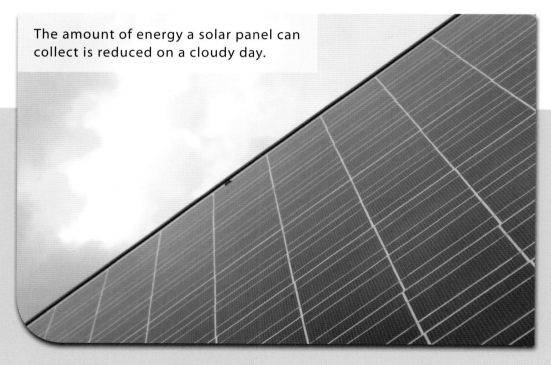

The amount of energy a solar panel can collect is reduced on a cloudy day.

Nighttime is also a problem. Solar systems can collect energy only during the day. And the amount of solar energy that is available changes. It all depends on the time of day, the time of year, the location, and the weather.

Some solar panels are designed to move according to the Sun's position. These panels collect as much energy as possible every day.

Sun Tracking

If you had animals on your farm, you'd want to keep track of them. The same goes for the Sun. Knowing how it moves in the sky can help you get the most energy out of it.

Consider a single spot on Earth. The energy that hits that spot is greatest when the Sun is highest in the sky. That happens right around midday. And summer days deliver more energy than winter days do. That's because Earth tilts on its axis. When it's summer in the Northern Hemisphere, the top half of Earth tilts toward the Sun. Sunlight is more direct then. And weather is usually warmer.

In the afternoon, the Sun is lower in the sky in the winter than in the summer.

During the winter, Earth's Northern Hemisphere points away from the Sun. The Sun stays lower in the sky. Its energy is spread out over a greater area. Shadows are longer. And there are fewer hours of daylight.

How will seasons affect the amount of energy a solar farm gathers?

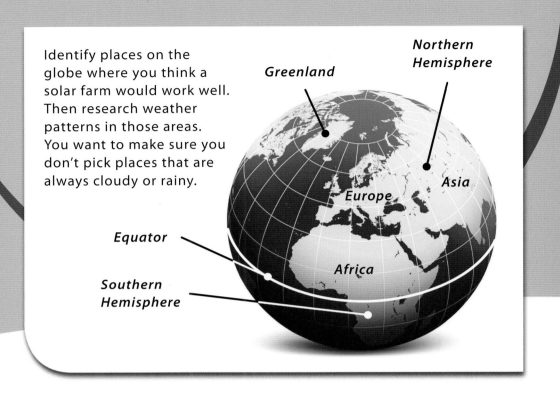

Identify places on the globe where you think a solar farm would work well. Then research weather patterns in those areas. You want to make sure you don't pick places that are always cloudy or rainy.

Greenland

Northern Hemisphere

Asia

Europe

Equator

Africa

Southern Hemisphere

From North to South

Seasons are a part of life in most places. Winter becomes spring and spring becomes summer as Earth **orbits** the Sun. Because of Earth's tilt, seasons are opposite on the two hemispheres. When it is winter in the United States, it is summer in Australia.

Over the course of a year, the equator gets the most solar energy. Areas closer to the poles get the least. In places that are really far north or south, the Sun may barely rise for months at a time.

What are some good places on Earth to build a solar farm? Think about which areas get the most direct sunlight for the most days of the year.

Solar energy enters a greenhouse, like this one, through the windows and heats up the air inside. The windows block heat energy from escaping back outside.

Fuel Effects

One reason to go solar is that it can keep the planet from getting too hot. When we burn fuels to heat buildings and power cars, we put lots of gases into the air. Carbon dioxide (CO_2) is a major one. It is called a greenhouse gas because it traps solar energy in the atmosphere. It works the same way that windows trap heat energy inside a greenhouse.

Greenhouse gases trap solar energy near Earth's surface. As a result, Earth's atmosphere gets warmer. This is called the **greenhouse effect.**

The greenhouse effect is important for our planet. Without it, heat energy would escape into space. And Earth would be too cold for people to survive. But too much warming can alter weather patterns and harm living things around the globe.

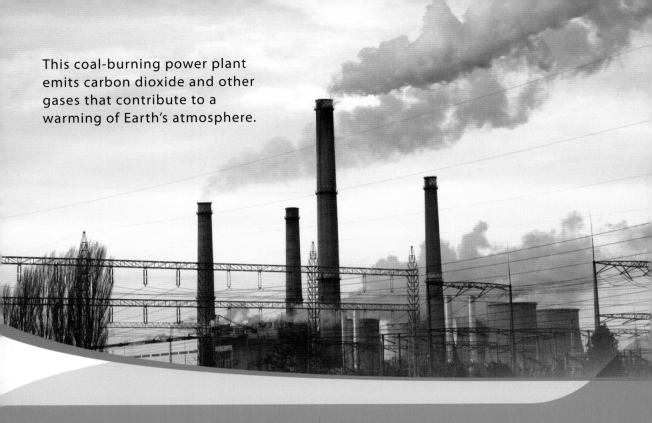

This coal-burning power plant emits carbon dioxide and other gases that contribute to a warming of Earth's atmosphere.

Levels of greenhouse gases in Earth's atmosphere have risen over the past few hundred years. And they continue to rise. For example, in 1960, the concentration of atmospheric CO_2 was about 317 parts per million (ppm). In 2010, the concentration was about 391 ppm. In turn, average temperatures have risen around the globe.

More warming could forever change Earth as we know it. Huge glaciers of snow and ice will melt. All of that melted water will flow into the sea. As sea levels rise, coastal areas will flood. Some areas on Earth will get hotter. Some might get colder, wetter, or drier. No one knows how plants and animals will fare as their environments change.

How can a solar farm help reduce greenhouse gases in the atmosphere?

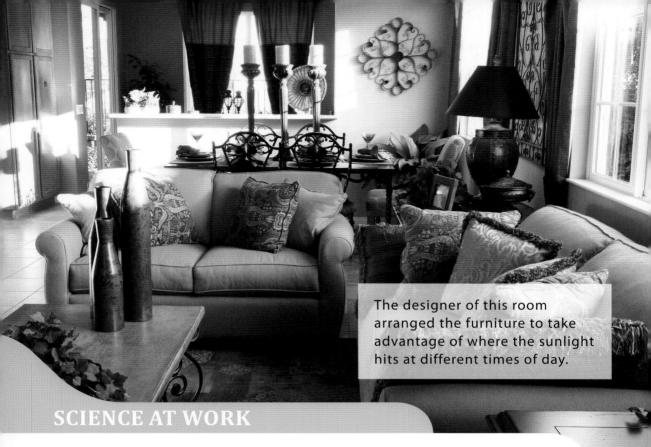

The designer of this room arranged the furniture to take advantage of where the sunlight hits at different times of day.

Interior Designer

Interior designers decorate offices, rooms, and homes. They pick out furniture, lamps, and rugs. Then, they put everything in its place. Sometimes they help decide where windows, stairs, and halls will go. Good designers know how color and light affect the temperature of a room. They think about the climate where they work. And they pick fabrics and materials that will make a room feel warmer or cooler. Designers also plan around windows. They want light and heat energy to flow well from one room to another.

People have always relied on the Sun for light and heat energy. Our ancestors used solar energy to dry food, clothes, clay, and other objects.

Using the Sun as a clothes dryer keeps down the demand for electricity. That way, power plants burn less coal.

Over time, people have found more and more ways to use solar energy. Today, some see the Sun as a better alternative to fossil fuels. Look around you. How many ways do you use the Sun to help you in your life?

If we used up all the oil and coal on Earth, there would be no more. Gas stations would become useless.

Use It or Lose It

Solar energy is a type of **renewable energy.** As long as Earth orbits the Sun, the energy will never run out. The Sun will continue to burn for billions of years.

Fossil fuels, on the other hand, are not renewable. Once we use them, they are gone forever.

Wind turbines help supply electricity for a canola farm.
Seeds from the canola plant are used to make cooking oil.

Wind is another source of renewable energy. It can power windmills. Windmills, also called wind turbines, are like huge pinwheels. They have large blades that spin in the wind. The blades turn a generator. The generator makes electricity. Many farms get electricity from **wind energy.**

Biofuels are another type of renewable energy. One example is alcohol made from corn. Corn can be planted and grown every year.

In Iceland's Blue Lagoon, the water is warm enough for people to swim year-round. The heat energy that warms the water comes from deep inside Earth. The same heat energy supplies energy to the power plant in the background.

Geothermal energy is yet another kind of renewable energy. It taps into heat energy that comes from inside Earth. Iceland uses a lot of geothermal energy. Steam seeps out of the ground in this island country in the North Atlantic Ocean. People use it to heat their homes.

Can you think of other types of renewable energy? How might you use these kinds of energy to help your solar farm?

Using solar energy leaves the environment clean so these daisies can thrive.

 Did You Know?

In 2007, NASA launched a new kind of spacecraft. Called *Dawn*, this craft used conventional rockets to escape Earth's atmosphere. Once in space, it began relying primarily on solar power to continue its journey to the asteroid belt. *Dawn's* solar arrays, which are 65 feet (20 meters) long, will collect enough energy to keep the spacecraft traveling and sending data back to Earth for eight years.

Why Solar?

You've learned a few reasons why it's good to use solar energy. For one thing, it is a clean energy. It does not pollute. And it does not emit greenhouse gases. Solar energy is also plentiful. It is available to everyone on Earth. No one country or company can control it. Fossil fuels, on the other hand, are more plentiful in some places than in others.

People have been making solar cars for years. But these autos aren't very practical. So far, no one has figured out how to make them work in bad weather!

Why Not Solar?

It's time for a reality check. Solar energy cannot solve all the world's problems. For one thing, we need light, power, and heat energy long after nightfall, and the Sun shines only during the day. Even then, clouds and rain can block the energy that is available. Also, it costs a lot to make electricity from solar energy. Compared to burning coal, it can cost twice as much.

Think about the iScience Puzzle. You need to lower the energy bills for a whole school. Can you think of ways to combine renewable energy with fossil fuels?

Sun Health

As you build your solar farm, you might spend a lot of time outside. That can be good and bad. Exposure to sunlight helps our bodies make **vitamin D.** Along with calcium, this vitamin helps your body build strong bones and teeth. It may also help keep your immune system healthy. And it can help protect against lots of diseases.

Be sure to cover up and protect your skin with sunscreen whenever you are going to be out in the sunlight.

But too much Sun exposure can damage skin. Sunburn, freckles, wrinkles, and tans are all signs of damage. **Skin cancer** is also a risk. People of all colors should be careful not to get too much sunlight.

As you build your farm, wear sunscreen. A sunscreen with a high SPF (Sun Protection Factor) provides more protection from the Sun than one with a low SPF. Put on a rimmed hat and long sleeves. And stay in the shade during the hours when the Sun is highest in the sky.

Throughout this book, you have learned why solar energy is good. You have seen some of its downsides. And you have learned a few of the ways you can catch energy from the Sun. Now, it's time to solve your school's problem. Look at your original plan. Do you want to make any changes to it? Here are some questions the school board will want you to answer.

How will you collect and use solar energy?

You could combine solar cells into arrays. Or you might use mirrors to collect and concentrate sunlight. You could then use warmed water to heat the school. Or you could make electricity for the light system. Which idea would work best where you live?

If we had lots more solar farms like this one, we could significantly reduce our need for fossil fuels.

Where should you build the farm?

Look at the environment around your school. You want to put your farm in a place without many hills or trees. That way, shadows won't block your sunlight. Do you live near the equator or closer to one of the poles? Is your area usually sunny or cloudy? Based on your answers, will you want to combine a solar farm with other sources of energy?

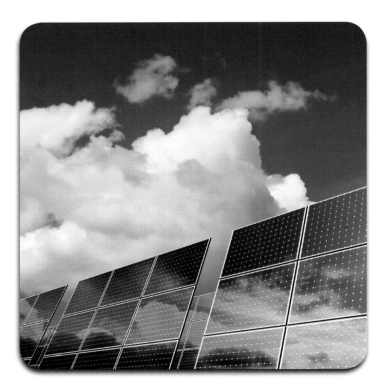

A location where the Sun shines many hours every day, every month of the year, would be a good spot for this solar array.

What materials will you use?

Think back to the Discover Activity. What kinds of materials will you use to get the most out of the Sun? What color will they be? How much do they cost? Will your total cost be less than the current bill?

Why is a solar farm useful?

This is your chance to sell the idea to the school board! List the environmental benefits. Explain how your plan can save money. There is a lot to consider when planning a solar farm. But if you plan it well and your reasons are good, you could make it happen!

This engineer is collecting data to make sure the solar panels are working as efficiently as intended.

There is no perfect system for collecting the Sun's energy on Earth. One reason is that our atmosphere blocks a lot of the energy that comes our way. Can you design a system to gather even more solar energy?

One strategy is to put solar arrays on satellites. These spacecraft orbit above the atmosphere. Can you think of other ways to collect solar energy in space? How will you transport it to Earth? Look on the Internet for ideas that scientists are already working on. Then, design your own system!

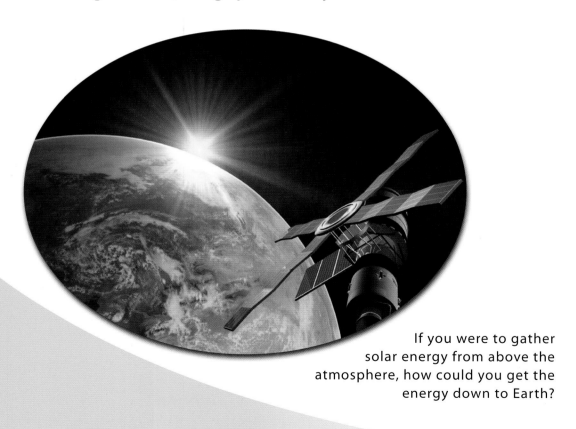

If you were to gather solar energy from above the atmosphere, how could you get the energy down to Earth?

GLOSSARY

absorb: to take in, as a sponge takes in water.

active solar heating: type of solar heating that uses special equipment to collect solar energy and change it to heat energy or electrical energy.

atoms: the basic units of matter.

biofuels: fuels made from plants.

concentrated: made stronger.

electric circuit: the path electricity follows.

electric current: the movement of electrons from one atom to the next in an electric circuit.

electricity: the movement of charged particles, usually electrons.

evaporate: to change from a liquid into a gas.

fossil fuels: coal, oil, and gas, which began forming in Earth's crust many millions of years ago.

geothermal energy: heat energy that comes from inside Earth.

greenhouse effect: the rise in temperature of Earth's atmosphere when gases in the atmosphere trap the Sun's energy.

helium: the gas that, along with hydrogen, makes up most of the Sun.

hydrogen: the gas that, along with helium, makes up most of the Sun.

nuclear fusion: process in which atoms join and release energy.

orbits: circles.

passive solar heating: type of solar heating that uses no special equipment to collect solar energy.

photosynthesis: process by which plants use the energy of the Sun to make food.

reflect: to reverse the direction of light or of an image, as a mirror does.

renewable energy: source of energy that will not run out.

silicon: common element used to make solar cells.

skin cancer: a disease of skin cells.

solar array: a group of connected solar modules.

solar cells: devices that turn solar energy into electrical energy.

solar collectors: devices that capture solar energy.

solar energy: energy from the Sun.

solar module: a group of connected solar cells.

vitamin D: vitamin that helps the human body absorb calcium.

water cycle: water passing from water vapor in the atmosphere, to rain or snow falling on Earth, and back up to the atmosphere as vapor again.

watts: the units in which electrical energy is measured.

wind energy: energy from moving air that turns windmills to operate machines or produce electricity.

FURTHER READING

Solar Power (Energy for Today), by Tea Benduhn.
Weekly Reader Early Learning Library, 2008.

Solar Cell and Renewable Energy Experiments, by Ed Sobey.
Enslow Publishers, 2011.

NASA, Sun for Kids.
http://www.nasa.gov/vision/universe/solarsystem/sun_for_kids_main.html

U.S. Department of Energy, Energy Savers.
www.energysavers.gov/your_home/electricity/index.cfm/mytopic=10490

U.S. Energy Information Administration, Energy Kids.
http://www.eia.doe.gov/kids/energy.cfm?page=solar_home-basics

ADDITIONAL NOTES

Page 8: The Sun's energy could heat the water. The water temperature could get higher than what a goldfish could live in.

Page 9: If you concentrate the Sun's energy, you can capture more of it from a smaller area.

Page 11: The water in the bowl with the black fabric likely got the warmest, followed by the water in the white bowl and then the water in the foil bowl. Dark colors help collect solar energy. Shiny or reflective materials can focus solar energy away from an area. However, when positioned correctly, they can also focus solar energy toward an area.

Page 12: Darker colors absorb energy better than lighter colors do.

Page 14: Since shadowed areas get less of the Sun's energy than sunny areas do, the amount of solar energy captured would vary with the position of the Sun in the sky. Sunlight reflected away from the solar farm would not be captured.

Page 18: The energy will be used to light and heat the school, and to power electric pencil sharpeners, floor cleaners, computers, and other electronic devices. Caption question: The panels on the ground likely supply electricity to the neighboring homes, or to a large building in the neighborhood, such as a school.

Page 20: Build structures that are designed to attract solar energy in dark colors and those that are not in light colors. Caption question: A light-colored house would be better in a warm climate.

Page 22: Solar cells might be used to create energy to run lights and other electrical equipment.

Page 25: Caption question: A large solar array might supply electricity or hot water or both to a large office building, a hospital, an airport, or some other large facility.

Page 29: During winter months, when there is less direct sunlight and fewer hours of sunlight, the amount of energy produced will decrease. During summer months, the opposite will be true.

Page 30: Good places to build solar farms would be near the equator where there is a need for a lot of energy.

Page 32: Solar energy is clean and it produces no greenhouse gases. Using solar energy would reduce use of fossil fuels, which do produce greenhouse gases.

Page 37: Other sources of renewable energy, such as flowing water or the wind, could help you produce more energy than you could produce with just solar energy alone.

Page 39: Fossil fuels could be used to heat the school building and solar energy could be used to generate electricity for lighting.